Morden Analytical Techniques: A Student Guide

Volume 1

Author

Arin Bhattacharya

DEDICATION

THIS BOOK IS DEDICATED

TO

MY RESPECTED PARENTS

I ALSO THANK GOD, MY FELLOW MATES

&

ALL THE LOVERS OF PHARMACY WORLD

Acknowledgement

In any of scientific literally work many people work towards achievement of the desired goal i.e. publication of the research work in form book. I want to thanks my parents without *their blessings, support and vision I could not have reached this stage.*

Friendship is a treasured gift and fine friends are very few, my heartful thanks to my friend s that are always there for their support and positive views.

I would like to thanks the persons whom had criticize this project because it provided the positive impetus for completing the project in a more comprehensive and positive way.

Lastly it is only when someone writes a book that one realizes the true power of MSOffice software. It is simple- without this software this book would not be written.

Thank you Mr. Bill Gates and Microsoft Corp!

I strongly believe, "Success is blend of hard work and Destiny" and I think 'God' who have perpetually patronized me with consciousness and love to ladder the success. Thankful I ever remain....

Arin Bhattacharya

Table of Contents

Chapter 1

GC MS

GC-MS is composed of two major building blocks: the gas chromatograph and the mass spectrometer. The gas chromatograph utilizes a capillary column which depends on the column's dimensions (length, diameter, film thickness) as well as the phase properties (e.g. 5% phenyl polysiloxane). The difference in the chemical properties between different molecules in a mixture and their relative affinity for the stationary phase of the column will promote separation of the molecules as the sample travels the length of the column. The molecules are retained by the column and then elute (come off) from the column at different times (called the retention time), and this allows the mass spectrometer downstream to capture, ionize, accelerate, deflect, and detect the ionized molecules separately. The mass spectrometer does this by breaking each molecule into ionized fragments and detecting these fragments using their mass-to-charge ratio.

These two components, used together, allow a much finer degree of substance identification than either unit used separately. It is not possible to make an accurate identification of a particular molecule by gas chromatography or mass spectrometry alone. The mass spectrometry process normally requires a very pure sample while gas chromatography using a traditional detector (e.g. Flame ionization detector) cannot differentiate between multiple molecules that happen to take the same amount of time to travel through the column (*i.e.* have the same retention time), which results in two or more molecules that co-elute. Sometimes two different molecules can also have a similar pattern of ionized fragments in a mass spectrometer (mass spectrum). Combining the two processes reduces the possibility of error, as it is extremely unlikely that two different molecules will behave in the same way in both a gas chromatograph.

and a mass spectrometer. Therefore, when an identifying mass spectrum appears at a characteristic retention time in a GC-MS analysis, it typically increases certainty that the analyte of interest is in the sample.

Figure 1 GC MS Instrument

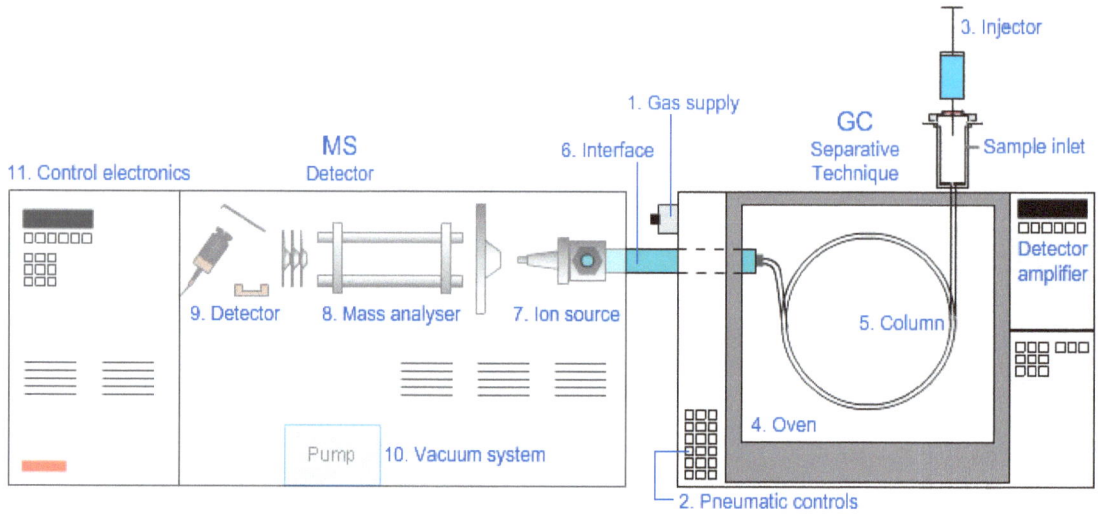

Figure 2 GC MS Instrument (Simplified diagram)

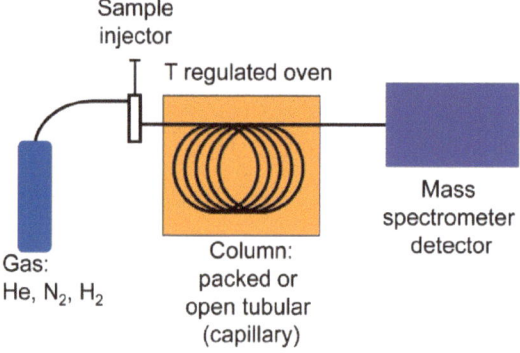

Purge and trap GC-MS

For the analysis of volatile compounds, a purge and trap (P&T) concentrator system may be used to introduce samples. The target analytes are extracted and mixed with water and introduced into an airtight chamber. An inert gas such as Nitrogen (N2) is bubbled through the water; this is known as purging. The volatile compounds move into the headspace above the water and are drawn along a pressure gradient (caused by the introduction of the purge gas) out of the chamber. The volatile compounds are drawn along a heated line onto a 'trap'. The trap is a column of adsorbent material at ambient temperature that holds the compounds by returning them to the liquid phase. The trap is then heated and the sample compounds are introduced to the GC-MS column via a volatiles interface, which is a split inlet system. P&T GC-MS is particularly suited to volatile organic compounds (VOCs) and BTEX compounds (aromatic compounds associated with petroleum).

Figure 3 Purge and trap GC-MS

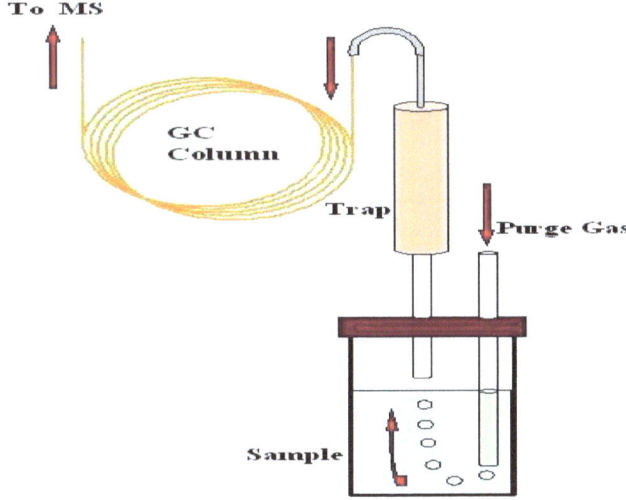

Types of mass spectrometer detectors

The most common type of mass spectrometer (MS) associated with a gas chromatograph (GC) is the quadrupole mass spectrometer, sometimes referred to by the Hewlett-Packard (now Agilent) trade name "Mass Selective Detector" (MSD). Another relatively common detector is the ion trap mass spectrometer. Additionally one may find a magnetic sector mass spectrometer, however these particular instruments are expensive and bulky and not typically found in high-throughput service laboratories. Other detectors may be encountered such as time of flight (TOF), tandem quadrupoles (MS-MS) (see below), or in the case of an ion trap MSn where n indicates the number mass spectrometry stages

GC-tandem MS

When a second phase of mass fragmentation is added, for example using a second quadrupole in a quadrupole instrument, it is called tandem MS (MS/MS). MS/MS can sometimes be used to quantitate low levels of target compounds in the presence of a high sample matrix background.

The first quadrupole (Q1) is connected with a collision cell (q2) and another quadrupole (Q3). Both quadrupoles can be used in scanning or static mode, depending on the type of MS/MS analysis being performed. Types of analysis include product ion scan, precursor ion scan, selected reaction monitoring (SRM) (sometimes referred to as multiple reaction monitoring (MRM)) and neutral loss scan. For example: When Q1 is in static mode (looking at one mass only as in SIM), and Q3 is in scanning mode, one obtains a so-called product ion spectrum (also called "daughter spectrum"). From this spectrum, one can select a prominent product ion which can be the product ion for the chosen precursor ion. The pair is called a "transition" and forms the basis for SRM. SRM is highly specific and virtually eliminates matrix background.

Ionization

After the molecules travel the length of the column, pass through the transfer line and enter into the mass spectrometer they are ionized by various methods with typically only one method being used at any given time. Once the sample is fragmented it will then be detected, usually by an electron multiplier diode, which essentially turns the ionized mass fragment into an electrical signal that is then detected.

The ionization technique chosen is independent of using full scan or SIM.

Electron ionization

By far the most common and perhaps standard form of ionization is electron ionization (EI). The molecules enter into the MS (the source is a quadrupole or the ion trap itself in an ion trap MS) where they are bombarded with free electrons emitted from a filament, not unlike the filament one would find in a standard light bulb. The electrons bombard the molecules, causing the molecule to fragment in a characteristic and reproducible way. This "hard ionization" technique results in the creation of more fragments of low mass to charge ratio (m/z) and few, if any, molecules approaching the molecular mass unit. Hard ionization is considered by mass spectrometrists as the employ of molecular electron bombardment, whereas "soft ionization" is charge by molecular collision with an introduced gas. The molecular fragmentation pattern is dependent upon the electron energy applied to the system, typically 70 eV (electron Volts).

Figure 4 Electron ionization

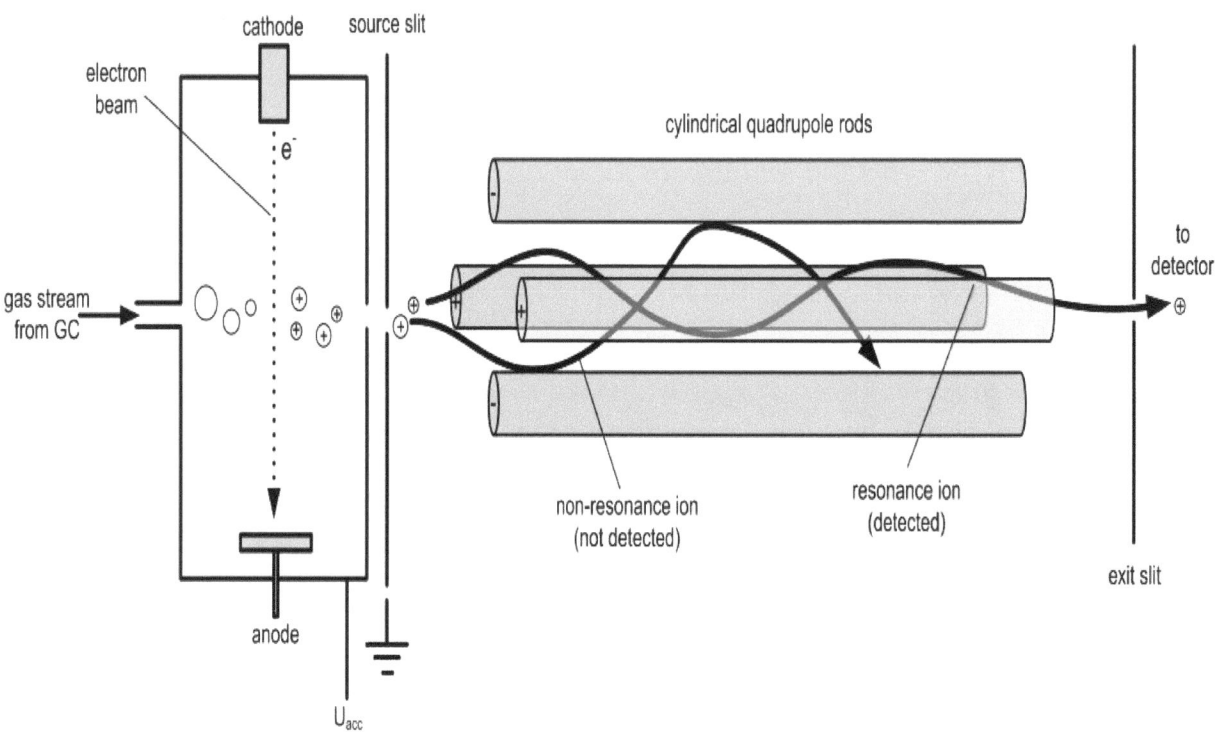

Cold electron ionization

The "hard ionization" process of electron ionization can be softened by the cooling of the molecules before their ionization, resulting in mass spectra that are richer in information. In this method named cold electron ionization (Cold-EI) the molecules exit the GC column, mixed with added helium make up gas and expand into vacuum through a specially designed supersonic nozzle, forming a supersonic molecular beam (SMB). Collisions with the makeup gas at the expanding supersonic jet reduce the internal vibrational (and rotational) energy of the analyte molecules, hence reducing the degree of fragmentation caused by the electrons during the ionization process. Cold-EI mass spectra are characterized by an abundant molecular ion while the usual fragmentation pattern is retained, thus making Cold-EI mass spectra compatible with library search identification techniques. The enhanced molecular ions increase the identification probabilities of both known and unknown compounds, amplify isomer mass spectral effects and enable the use of isotope abundance analysis for the elucidation of elemental formulae

Chemical ionization

In chemical ionization a reagent gas, typically methane or ammonia is introduced into the mass spectrometer. Depending on the technique (positive CI or negative CI) chosen, this reagent gas will interact with the electrons and analyte and cause a 'soft' ionization of the molecule of interest. A softer ionization fragments the molecule to a lower degree than the hard ionization of EI. One of the main benefits of using chemical ionization is that a mass fragment closely corresponding to the molecular weight of the analyte of interest is produced.

In positive chemical ionization (PCI) the reagent gas interacts with the target molecule, most often with a proton exchange. This produces the species in relatively high amounts.

In negative chemical ionization (NCI) the reagent gas decreases the impact of the free electrons on the target analyte. This decreased energy typically leaves the fragment in great supply.

Figure 6 Chemical ionization reactions

A) EI of reagent gas to form ions:

$$CH_4 + e^- \rightarrow CH_4^{\cdot+} + 2e^-$$

B) Reaction of reagent gas ions to form adducts:

$$CH_4^{\cdot+} + CH_4 \rightarrow CH_5^+ + CH_3\cdot$$

OR

$$CH_4^{\cdot+} \rightarrow CH_3^+ + H\cdot$$
$$CH_3^+ + CH_4 \rightarrow C_2H_5^+ + H_2$$

C) Reaction of Reagent Gas Ions with analyte molecules:

$$CH_5^+ + M \rightarrow CH_4 + MH^+$$

$$C_2H_5^+ + M \rightarrow C_2H_4 + MH^+$$

$$CH_5^+ + M \rightarrow CH_4 + (M-H)^+$$

Method of analysis

A mass spectrometer is typically utilized in one of two ways: full scan or selected ion monitoring (SIM). The typical GC-MS instrument is capable of performing both functions either individually or concomitantly, depending on the setup of the particular instrument.

The primary goal of instrument analysis is to quantify an amount of substance. This is done by comparing the relative concentrations among the atomic masses in the generated spectrum. Two kinds of analysis are possible, comparative and original. Comparative analysis essentially compares the given spectrum to a spectrum library to see if its characteristics are present for some sample in the library. This is best performed by a computer because there are a myriad of visual distortions that can take place due to variations in scale. Computers can also simultaneously correlate more data (such as the retention times identified by GC), to more accurately relate certain data.

Another method of analysis measures the peaks in relation to one another. In this method, the tallest peak is assigned 100% of the value, and the other peaks being assigned proportionate values. All values above 3% are assigned. The total mass of the unknown compound is normally indicated by the parent peak. The value of this parent peak can be used to fit with a chemical formula containing the various elements which are believed to be in the compound. The isotope pattern in the spectrum, which is unique for elements that have many isotopes, can also be used to identify the various elements present. Once a chemical formula has been matched to the spectrum, the molecular structure and bonding can be identified, and must be consistent with the characteristics recorded by GC-MS. Typically, this identification done automatically by programs which come with the instrument, given a list of the elements which could be present in the sample.

A "full spectrum" analysis considers all the "peaks" within a spectrum. Conversely, selective ion monitoring (SIM) only monitors selected ions associated with a specific substance. This is done on the assumption that at a given retention time, a set of ions is characteristic of a certain compound. This is a fast and efficient analysis, especially if the analyst has previous information about a sample or is only looking for a few specific substances. When the amount of information collected about the ions in a given gas chromatographic peak decreases, the sensitivity of the analysis increases. So, SIM analysis allows for a smaller quantity of a compound to be detected and measured, but the degree of certainty about the identity of that compound is reduced.

Full scan MS

When collecting data in the full scan mode, a target range of mass fragments is determined and put into the instrument's method. An example of a typical broad range of mass fragments to monitor would be m/z 50 to m/z 400. The determination of what range to use is largely dictated by what one anticipates being in the sample while being cognizant of the solvent and other possible interferences. A MS should not be set to look for mass fragments too low or else one may detect air (found as m/z 28 due to nitrogen), carbon dioxide (m/z 44) or other possible interferences. Additionally if one is to use a large scan range then sensitivity of the instrument is decreased due to performing fewer scans per second since each scan will have to detect a wide range of mass fragments.

Full scan is useful in determining unknown compounds in a sample. It provides more information than SIM when it comes to confirming or resolving compounds in a sample. During instrument method development it may be common to first analyze test solutions in full scan mode to determine the retention time and the mass fragment fingerprint before moving to a SIM instrument method.

Selected ion monitoring

In selected ion monitoring (SIM) certain ion fragments are entered into the instrument method and only those mass fragments are detected by the mass spectrometer. The advantages of SIM are that the detection limit is lower since the instrument is only looking at a small number of fragments (e.g. three fragments) during each scan. More scans can take place each second. Since only a few mass fragments of interest are being monitored, matrix interferences are typically lower. To additionally confirm the likelihood of a potentially positive result, it is relatively important to be sure that the ion ratios of the various mass fragments are comparable to a known reference standard

Applications of GC MS

1. Environmental monitoring and cleanup
2. Criminal forensics
3. Law enforcement
4. Sports anti-doping analysis
5. Security
6. Food, beverage and perfume analysis
7. Astrochemistry
8. Medicine

Specific applications of GC MS

Gas chromatography-mass spectrometry (GC-MS) methods for the analysis of amphetamines, ketamines, opioids, cocaine, and other abused drugs in urine.

Explanation using amphetamine

- Amphetamine (AM, *R, S-1-phenyl-2-propanamine)* and methamphetamine (MA, *R, S-N-methyl-1- phenyl-* 2-propanamine) are powerful stimulants in the central nervous system. These drugs are often abused and used doping agents in sports.
- Stimulants in doping control consists of a heterogeneous group of compounds, the majority of which is structurally related to amphetamine
- Structures of amphetamine

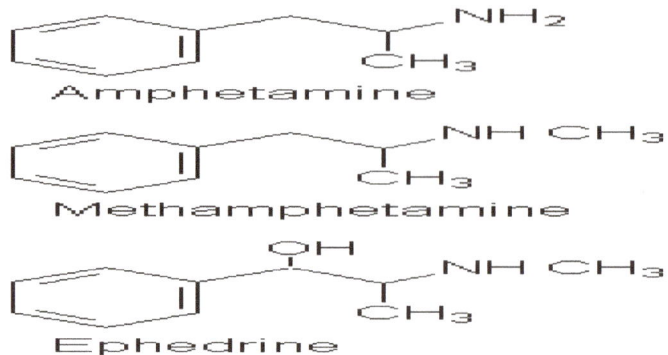

GC MS USED IN THE ANALYSIS

The Finnigan TRACE DSQ is used for analysis

Figure 7 The Finnigan TRACE DSQ

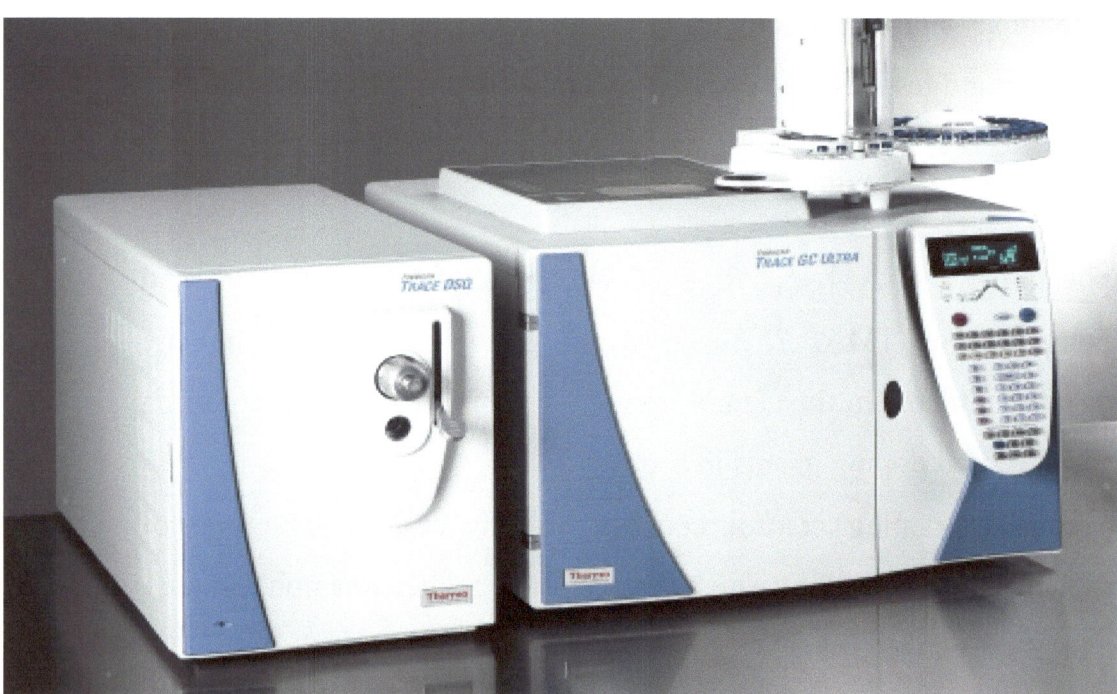

- Split/splitless injection port

- AS 3000 autosampler.

- Optional vacuum probe interlock

15 m x 0.25 mm i.d. x 0.25 μm TR-5MS column

- Oven temperature was held at 190°C

- Isothermal

- Total GC run time was 4.0 minutes

- 2.0 mL/min flow rate for carrier gas (He).

- Electron ionization

- Quadrupole mass spectrometer

- Uses SIM(Selected Ion Mode)

Figure 8 GC MS Spectra of Amphetamine using above mentioned instrument

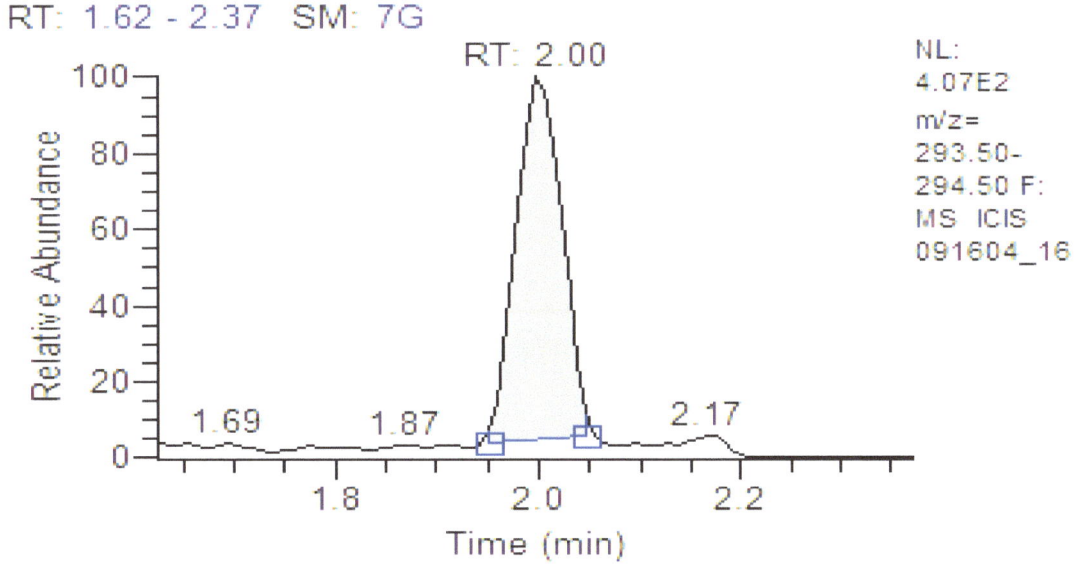

TR-5MS column details

TRACE TR-5MS Columns provide low bleed separations and are ideal for use in GC/MS applications, with a maximum operating temperature of 370°C. These inert columns result in minimal peak tailing and works well for environmental semi-volatile and food applications. The TR-5MS columns show excellent robustness and are the ideal choice for GC/MS instruments doing a wide variety of methods.

Specifications

Product Type	Capillary Column
Max temperature (° C)	370
Bonded Phase	5% Phenyl polysilphenylene-siloxane
Typical compounds	GC/MS applications
Film thickness (micron)	0.25

Limitations

Although many consider GC/MS to be the "gold standard" in scientific analysis, GC/MS does have some limitations. Because great faith is maintained in GC/MS analysis, erroneous results are not expected and hard to dispute. However, false positives and false negatives are possible.

Some problems with GC/MS originate in improper conditions in the GC portion of the analysis. If the GC instrument does not separate the specimen's compounds completely, the MS feed is impure. This usually results in background "noise" in the mass spectrum. If the carrier gas in the GC process is not correctly deflected from entering the MS instrument, similar contamination may occur.

Also, the MS portion suffers from the inexact practice of interpreting mass spectra. An analyst must correlate computer calculations with system conditions. The typical memory bank for MS identification contains about 5000 spectra for a particular group of compounds. Even if a competent analyst could find conclusive results pointing to one substance out of 5000 substances, this does not rule out the remaining over 200,000 known existing chemicals. For the 5000-spectra memory bank, the typical computer result is limited to as many as six possible identifications.

In one instance, erroneous GC/MS results may have been responsible for a criminal defendant receiving a death sentence. John Brown killed a police officer and wounded two bar patrons in a shoot-out on June 7, 1980 in Garden Grove, California. Mr. Brown's diminished capacity defense to capital murder relied on the assertion that Mr. Brown was under the influence of narcotics at the time of the shooting. The prosecution introduced GC/MS evidence that showed Mr. Brown's blood to be free of narcotics. The California Supreme Court overturned the jury's death sentence because the prosecution never introduced evidence from a radioactive immunoassay ("RIA") test that detected phencyclidine (PCP) in Mr. Brown's blood. Obviously, an example like this demonstrates that analytical evidence, including GC/MS, should always be confirmed with another reliable technique.

A more advanced analytical method is MS/MS, a tandem series of instruments, which has the advantage of increased sensitivity. One court states that MS/MS analysis has never produced a false positive in the FBI laboratory. However, MS/MS is not widely used yet as the instrument's cost is prohibitive.

Conclusion

GC and MS are useful tools for chemical analysis, especially when used together.

Chapter 2 The Transmission Electron Microscope

Introduction to Transmission Electron Microscope

The transmission electron microscope (TEM) operates on the same basic principles as the light microscope but uses electrons instead of light. What you can see with a light microscope is limited by the wavelength of light. TEMs use electrons as "light source" and their much lower wavelength makes it possible to get a resolution a thousand times better than with a light microscope.

You can see objects to the order of a few angstrom (10^{-10} m). For example, you can study small details in the cell or different materials down to near atomic levels. The possibility for high magnifications has made the TEM a valuable tool in both medical, biological and materials research.

A "light source" at the top of the microscope emits the electrons that travel through vacuum in the column of the microscope. Instead of glass lenses focusing the light in the light microscope, the TEM uses electromagnetic lenses to focus the electrons into a very thin beam. The electron beam then travels through the specimen you want to study. Depending on the density of the material present, some of the electrons are scattered and disappear from the beam. At the bottom of the microscope the unscattered electrons hit a fluorescent screen, which gives rise to a "shadow image" of the specimen with its different parts displayed in varied darkness according to their density. The image can be studied directly by the operator or photographed with a camera.

Figure 9 Transmission Electron Microscope Instrumentation

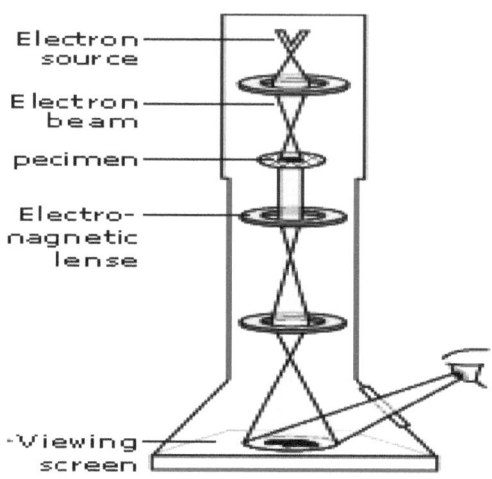

Background

Electrons

Theoretically, the maximum resolution, d, that one can obtain with a light microscope has been limited by the wavelength of the photons that are being used to probe the sample, λ and thenumerical aperture of the system,

$$d = \frac{\lambda}{2n\sin\alpha} \approx \frac{\lambda}{2\,\mathrm{NA}}$$

Early twentieth century scientists theorised ways of getting around the limitations of the relatively large wavelength of visible light (wavelengths of 400–700 nanometers) by using electrons. Like all matter, electrons have both wave and particle properties (as theorized by Louis-Victor de Broglie), and their wave-like properties mean that a beam of electrons can be made to behave like a beam of electromagnetic radiation. The wavelength of electrons is related to their kinetic energy via the de Broglie equation. An additional correction must be made to account for relativistic effects, as in a TEM an electron's velocity approaches the speed of light, c.

$$\lambda_e \approx \frac{h}{\sqrt{2m_0 E \left(1 + \frac{E}{2m_0 c^2}\right)}}$$

where, h is Planck's constant, m_0 is the rest mass of an electron and E is the energy of the accelerated electron. Electrons are usually generated in an electron microscope by a process known as thermionic emission from a filament, usually tungsten, in the same manner as a light bulb, or alternatively by field electron emission. The electrons are then accelerated by an electric potential (measured in volts) and focused by electrostatic and electromagnetic lenses onto the sample. The transmitted beam contains information about electron density, phase and periodicity; this beam is used to form an image.

Source formation

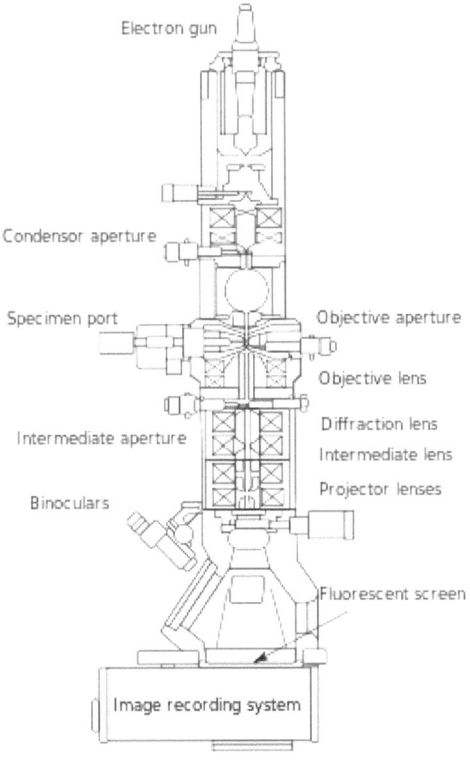

Figure 10 Layout of optical components in a basic TEM

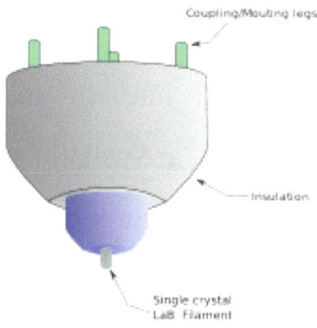

Coupling/Mouting legs

Insulation

Single crystal
LaB filament

Figure 11 Single crystal LaB$_6$ filament

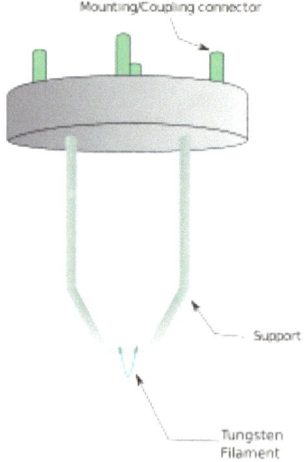

Mounting/Coupling connector

Support

Tungsten
Filament

From the top down, the TEM consists of an emission source, which may be a tungsten filament, or a lanthanum hexaboride (LaB$_6$) source For tungsten, this will be of the form of either a hairpin-style filament, or a small spike-shaped filament. LaB$_6$ sources utilize small single crystals. By connecting this gun to a high voltage source (typically ~100–300 kV) the gun will, given sufficient current, begin to emit electrons either by thermionicor field electron emission into the vacuum. This extraction is usually aided by the use of a Wehnelt cylinder. Once extracted, the upper lenses of the TEM allow for the formation of the electron probe to the desired size and location for later interaction with the sample. Manipulation of the electron beam is performed using two physical effects. The interaction of electrons with a magnetic field will cause electrons to move according to the right hand rule, thus allowing for electromagnets to manipulate the electron beam. The use of magnetic fields allows for the formation of a magnetic lens of variable focusing power, the lens shape originating due to the distribution of magnetic flux. Additionally, electrostatic fields can cause the electrons to be deflected through a constant angle. Coupling of two deflections in opposing directions with a small intermediate gap allows for the formation of a shift in the beam path, this being used in TEM for beam shifting, subsequently this is extremely important to STEM. From these two effects, as well as the use of an electron imaging system, sufficient control over the beam path is possible for TEM operation. The optical configuration of a TEM can be rapidly changed, unlike that for an optical microscope, as lenses in the beam path can be enabled, have their strength changed, or be disabled entirely simply via rapid electrical switching, the speed of which is limited by effects such as the magnetic hysteresis of the lenses. Lens that are being used to probe the sample, λ and the numerical aperture of the system, NA.

$$d = \frac{\lambda}{2n\sin\alpha} \approx \frac{\lambda}{2\,\text{NA}}$$

Early twentieth century scientists theorised ways of getting around the limitations of the relatively large wavelength of visible light (wavelengths of 400–700 nanometers) by using electrons. Like all matter, electrons have both wave and particle properties (as theorized by Louis-Victor de Broglie), and their wave-like properties mean that a beam of electrons can be made to behave like a beam of electromagnetic radiation. The wavelength of electrons is related to their kinetic energy via the de Broglie equation. An additional correction must be made to account for relativistic effects, as in a TEM an electron's velocity approaches the speed of light, c.

$$\lambda_e \approx \frac{h}{\sqrt{2m_0 E \left(1 + \frac{E}{2m_0 c^2}\right)}}$$

where, h is Planck's constant, m_0 is the rest mass of an electron and E is the energy of the accelerated electron. Electrons are usually generated in an electron microscope by a process known as thermionic emission from a filament, usually tungsten, in the same manner as a light bulb, or alternatively by field electron emission. The electrons are then accelerated by an electric potential (measured in volts) and focused by electrostatic and electromagnetic lenses onto the sample. The transmitted beam contains information about electron density, phase and periodicity; this beam is used to form an image.

A TEM is composed of several components, which include a vacuum system in which the electrons travel, an electron emission source for generation of the electron stream, a series of electromagnetic lenses, as well as electrostatic plates. The latter two allow the operator to guide and manipulate the beam as required. Also required is a device to allow the insertion into, motion within, and removal of specimens from the beam path. Imaging devices are subsequently used to create an image from the electrons that exit the system.

Vacuum system

To increase the mean free path of the electron gas interaction, a standard TEM is evacuated to low pressures, typically on the order of 10^{-4} Pa The need for this is twofold: first the allowance for the voltage difference between the cathode and the ground without generating an arc, and secondly to reduce the collision frequency of electrons with gas atoms to negligible levels—this effect is characterised by the mean free path. TEM components such as specimen holders and film cartridges must be routinely inserted or replaced requiring a system with the ability to re-evacuate on a regular basis. As such, TEMs are equipped with multiple pumping systems and airlocks and are not permanently vacuum sealed.

The vacuum system for evacuating a TEM to an operating pressure level consists of several stages. Initially a low or roughing vacuum is achieved with either a rotary vane pump or diaphragm pumps bringing the TEM to a sufficiently low pressure to allow the operation of a turbomolecular or diffusion pump which brings the TEM to its high vacuum level necessary for operations. To allow for the low vacuum pump to not require continuous operation, while continually operating the turbomolecular pumps, the vacuum side of a low-pressure pump may be connected to chambers which accommodate the exhaust gases from the turbomolecular pump. Sections of the TEM may be isolated by the use of pressure-limiting apertures, to allow for different vacuum levels in specific areas, such as a higher vacuum of 10^{-4} to 10^{-7} Pa or higher in the electron gun in high-resolution or field-emission TEMs.

High-voltage TEMs require ultra-high vacuums on the range of 10^{-7} to 10^{-9} Pa to prevent generation of an electrical arc, particularly at the TEM cathode.[25] As such for higher voltage TEMs a third vacuum system may operate, with the gun isolated from the main chamber either by use of gate valves or by the use of a differential pumping aperture. The differential pumping aperture is a small hole that prevents diffusion of gas molecules into the higher vacuum gun area faster than they can be pumped out. For these very low pressures either an ion pump or a getter material is used.

Poor vacuum in a TEM can cause several problems, from deposition of gas inside the TEM onto the specimen as it is being viewed through a process known as electron beam induced deposition, or in more severe cases damage to the cathode from an electrical discharge. Vacuum problems due to specimen sublimation are limited by the use of a cold trap to adsorb sublimated gases in the vicinity of the specimen.

Specimen stage

Figure 12 TEM sample support mesh "grid", with ultramicrotomy sections

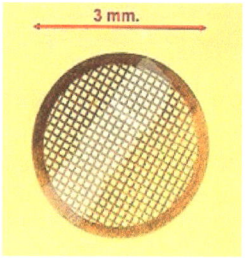

TEM specimen stage designs include airlocks to allow for insertion of the specimen holder into the vacuum with minimal increase in pressure in other areas of the microscope. The specimen holders are adapted to hold a standard size of grid upon which the sample is placed or a standard size of self-supporting specimen. Standard TEM grid sizes are a 3.05 mm diameter ring, with a thickness and mesh size ranging from a few to 100 μm. The sample is placed onto the inner meshed area having diameter of approximately 2.5 mm. Usual grid materials are copper, molybdenum, gold or platinum. This grid is placed into the sample holder, which is paired with the specimen stage. A wide variety of designs of stages and holders exist, depending upon the type of experiment being performed. In addition to 3.05 mm grids, 2.3 mm grids are sometimes, if rarely, used. These grids were particularly used in the mineral sciences where a large degree of tilt can be required and where specimen material may be extremely rare. Electron transparent specimens have a thickness around 100 nm, but this value depends on the accelerating voltage.

Once inserted into a TEM, the sample often has to be manipulated to present the region of interest to the beam, such as in single grain diffraction, in a specific orientation. To accommodate this, the TEM stage includes mechanisms for the translation of the sample in the XY plane of the sample, for Z height adjustment of the sample holder, and usually for at least one rotation degree of freedom for the sample. Thus a TEM stage may provide four degrees of freedom for the motion of the specimen. Most modern TEMs provide the ability for two orthogonal rotation angles of movement with specialized holder designs called double-tilt sample holders. Of note however is that some stage designs, such as top-entry or vertical insertion stages once common for high resolution TEM studies, may simply only have X-Y translation available. The design criteria of TEM stages are complex, owing to the simultaneous requirements of mechanical and electron-optical constraints and have thus generated many unique implementations.

A TEM stage is required to have the ability to hold a specimen and be manipulated to bring the region of interest into the path of the electron beam. As the TEM can operate over a wide range of magnifications, the stage must simultaneously be highly resistant to mechanical drift, with drift requirements as low as a few nm/minute while being able to move several μm/minute, with repositioning accuracy on the order of nanometers. Earlier designs of TEM accomplished this with a complex set of mechanical downgearing devices, allowing the operator to finely control the motion of the stage by several rotating rods. Modern devices may use electrical stage designs, using screw gearing in concert with stepper motors, providing the operator with a computer-based stage input, such as a joystick or trackball.

Two main designs for stages in a TEM exist, the side-entry and top entry version. Each design must accommodate the matching holder to allow for specimen insertion without either damaging delicate TEM optics or allowing gas into TEM systems under vacuum.

The most common is the side entry holder, where the specimen is placed near the tip of a long metal (brass or stainless steel) rod, with the specimen placed flat in a small bore. Along the rod are several polymer vacuum rings to allow for the formation of a vacuum seal of sufficient quality, when inserted into the stage. The stage is thus designed to accommodate the rod, placing the sample either in between or near the objective lens, dependent upon the objective design. When inserted into the stage, the side entry holder has its tip contained within the TEM vacuum, and the base is presented to atmosphere, the airlock formed by the vacuum rings.

Insertion procedures for side-entry TEM holders typically involve the rotation of the sample to trigger micro switches that initiate evacuation of the airlock before the sample is inserted into the TEM column.

The second design is the top-entry holder consists of a cartridge that is several cm long with a bore drilled down the cartridge axis. The specimen is loaded into the bore, possibly utilising a small screw ring to hold the sample in place. This cartridge is inserted into an airlock with the bore perpendicular to the TEM optic axis. When sealed, the airlock is manipulated to push the cartridge such that the cartridge falls into place, where the bore hole becomes aligned with the beam axis, such that the beam travels down the cartridge bore and into the specimen. Such designs are typically unable to be tilted without blocking the beam path or interfering with the objective lens.

Electron gun

<u>Figure 13 Electron gun</u>

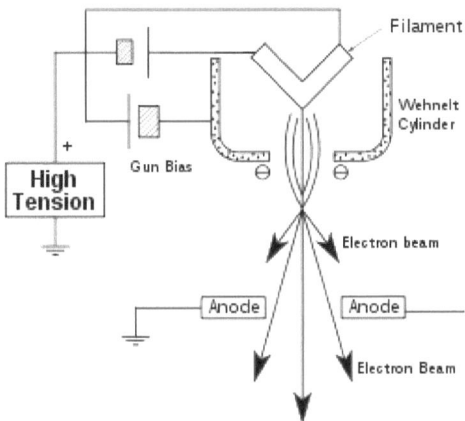

The electron gun is formed from several components: the filament, a biasing circuit, a Wehnelt cap, and an extraction anode. By connecting the filament to the negative component power supply, electrons can be "pumped" from the electron gun to the anode plate, and TEM column, thus completing the circuit. The gun is designed to create a beam of electrons exiting from the assembly at some given angle, known as the gun divergence semiangle, α. By constructing the Wehnelt cylinder such that it has a higher negative charge than the filament itself, electrons that exit the filament in a diverging manner are, under proper operation, forced into a converging pattern the minimum size of which is the gun crossover diameter.

The thermionic emission current density, J, can be related to the work function of the emitting material and is a Boltzmann distribution given below, where A is a constant, Φ is the work function and T is the temperature of the material.

$$J = AT^2 \exp\left(-\frac{\Phi}{kT}\right)$$

This equation shows that in order to achieve sufficient current density it is necessary to heat the emitter, taking care not to cause damage by application of excessive heat, for this reason materials with either a high melting point, such as tungsten, or those with a low work function (LaB_6) are required for the gun filament. Furthermore both lanthanum hexaboride and tungsten thermionic sources must be heated in order to achieve thermionic emission, this can be achieved by the use of a small resistive strip. To prevent thermal shock, there is often a delay enforced in the application of current to the tip, to prevent thermal gradients from damaging the filament, the delay is usually a few seconds for LaB_6, and significantly lower for tungsten[

Electron lens

Electron lenses are designed to act in a manner emulating that of an optical lens, by focusing parallel rays at some constant focal length. Lenses may operate electrostatically or magnetically. The majority of electron lenses for TEM utilise electromagnetic coils to generate a convex lens. For these lenses the field produced for the lens must be radially symmetric, as deviation from the radial symmetry of the magnetic lens causes aberrations such as astigmatism, and worsens spherical and chromatic aberration. Electron lenses are manufactured from iron, iron-cobalt or nickel cobalt alloyssuch as permalloy. These are selected for their magnetic properties, such as magnetic saturation, hysteresis and permeability.

The components include the yoke, the magnetic coil, the poles, the polepiece, and the external control circuitry. The polepiece must be manufactured in a very symmetrical manner, as this provides the boundary conditions for the magnetic field that forms the lens. Imperfections in the manufacture of the polepiece can induce severe distortions in the magnetic field symmetry, which induce distortions that will ultimately limit the lenses' ability to reproduce the object plane. The exact dimensions of the gap, pole piece internal diameter and taper, as well as the overall design of the lens is often performed by finite element analysis of the magnetic field, whilst considering the thermal and electrical constraints of the design.

The coils which produce the magnetic field are located within the lens yoke. The coils can contain a variable current, but typically utilise high voltages, and therefore require significant insulation in order to prevent short-circuiting the lens components. Thermal distributors are placed to ensure the extraction of the heat generated by the energy lost to resistance of the coil windings. The windings may be water-cooled, using a chilled water supply in order to facilitate the removal of the high thermal duty.

Apertures

Apertures are annular metallic plates, through which electrons that are further than a fixed distance from the optic axis may be excluded. These consist of a small metallic disc that is sufficiently thick to prevent electrons from passing through the disc, whilst permitting axial electrons. This permission of central electrons in a TEM causes two effects simultaneously: firstly, apertures decrease the beam intensity as electrons are filtered from the beam, which may be desired in the case of beam sensitive samples. Secondly, this filtering removes electrons that are scattered to high angles, which may be due to unwanted processes such as spherical or chromatic aberration, or due to diffraction from interaction within the sample. Apertures are either a fixed aperture within the column, such as at the condensor lens, or are a movable aperture, which can be inserted or withdrawn from the beam path, or moved in the plane perpendicular to the beam path. Aperture assemblies are mechanical devices which allow for the selection of different aperture sizes, which may be used by the operator to trade off intensity and the filtering effect of the aperture. Aperture assemblies are often equipped with micrometers to move the aperture, required during optical calibration.

Imaging methods

Imaging methods in TEM utilize the information contained in the electron waves exiting from the sample to form an image. The projector lenses allow for the correct positioning of this electron wave distribution onto the viewing system. The observed intensity of the image, I, assuming sufficiently high quality of imaging device, can be approximated as proportional to the time-average amplitude of the electron wavefunctions, where the wave which form the exit beam is denoted by Ψ.

$$I(x) = \frac{k}{t_1 - t_0} \int_{t_0}^{t_1} \Psi\Psi^* \, dt$$

Different imaging methods therefore attempt to modify the electron waves exiting the sample in a form that is useful to obtain information with regards to the sample, or beam itself. From the previous equation, it can be deduced that the observed image depends not only on the amplitude of beam, but also on the phase of the electrons, although phase effects may often be ignored at lower magnifications. Higher resolution imaging requires thinner samples and higher energies of incident electrons. Therefore the sample can no longer be considered to be absorbing electrons, via a Beer's law effect, rather the sample can be modelled as an object that does not change the amplitude of the incoming electron wavefunction. Rather the sample modifies the phase of the incoming wave; this model is known as a pure phase object, for sufficiently thin specimens phase effects dominate the image, complicating analysis of the observed intensities. For example, to improve the contrast in the image the TEM may be operated at a slight defocus to enhance contrast, owing to convolution by the contrast transfer function of the TEM, which would normally decrease contrast if the sample was not a weak phase object.

Contrast formation

Contrast formation in the TEM depends greatly on the mode of operation. Complex imaging techniques, which utilise the unique ability to change lens strength or to deactivate a lens, allow for many operating modes. These modes may be used to discern information that is of particular interest to the investigator.

Bright field

The most common mode of operation for a TEM is the bright field imaging mode. In this mode the contrast formation, when considered classically, is formed directly by occlusion and absorption of electrons in the sample. Thicker regions of the sample, or regions with a higher atomic number will appear dark, whilst regions with no sample in the beam path will appear bright – hence the term "bright field". The image is in effect assumed to be a simple two dimensional projection of the sample down the optic axis, and to a first approximation may be modelled via Beer's law more complex analyses require the modelling of the sample to include phase information.

Diffraction contrast

Samples can exhibit diffraction contrast, whereby the electron beam undergoes Bragg scattering, which in the case of a crystalline sample, disperses electrons into discrete locations in the back focal plane. By the placement of apertures in the back focal plane, i.e. the objective aperture, the desired Bragg reflections can be selected (or excluded), thus only parts of the sample that are causing the electrons to scatter to the selected reflections will end up projected onto the imaging apparatus.

If the reflections that are selected do not include the unscattered beam (which will appear up at the focal point of the lens), then the image will appear dark wherever no sample scattering to the selected peak is present, as such a region without a specimen will appear dark. This is known as a dark-field image.

Modern TEMs are often equipped with specimen holders that allow the user to tilt the specimen to a range of angles in order to obtain specific diffraction conditions, and apertures placed above the specimen allow the user to select electrons that would otherwise be diffracted in a particular direction from entering the specimen.

Applications

Applications for this method include the identification of lattice defects in crystals. By carefully selecting the orientation of the sample, it is possible not just to determine the position of defects but also to determine the type of defect present. If the sample is oriented so that one particular plane is only slightly tilted away from the strongest diffracting angle (known as the Bragg Angle), any distortion of the crystal plane that locally tilts the plane to the Bragg angle will produce particularly strong contrast variations. However, defects that produce only displacement of atoms that do not tilt the crystal to the Bragg angle (i. e. displacements parallel to the crystal plane) will not produce strong contrast.

Chapter 3 Differential scanning calorimetry

Differential scanning calorimetry is a technique we use to study what happens to polymers when they're heated. We use it to study what we call the *thermal transitions* of a polymer. And what are thermal transitions? They're the changes that take place in a polymer when you heat it. The melting of a crystalline polymer is one example. The glass transition is also a thermal transition.

So how do we study what happens to a polymer when we heat it? The first step would be to heat it, obviously. And that's what we do in differential scanning calorimetry, or DSC for short.

We heat our polymer in a device that looks something like this:

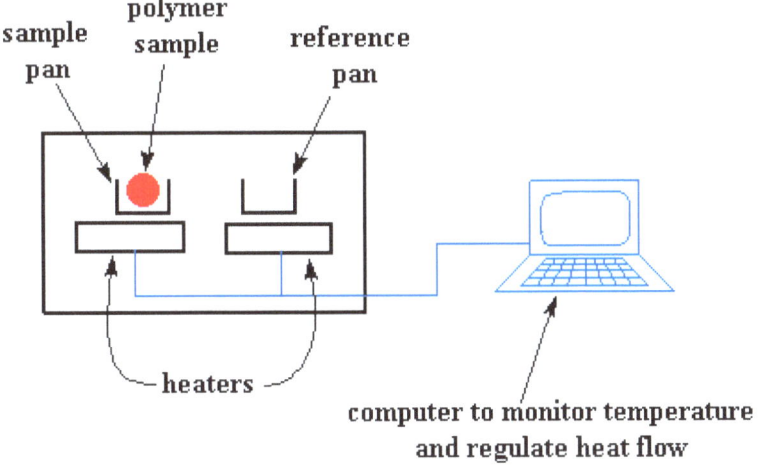

It's pretty simple, really. There are two pans. In one pan, the sample pan, you put your polymer sample. The other one is the reference pan. You leave it empty. Each pan sits on top of a heater. Then you tell the nifty computer to turn on the heaters. So the computer turns on the heaters, and tells it to heat the two pans at a specific rate, usually something like 10 °C per minute. The computer makes absolutely sure that the heating the rate stays exactly the same throughout the experiment.

But more importantly, it makes sure that the two separate pans, with their two separate heaters, heat at the same rate as each other.

Huh? Why wouldn't they heat at the same rate? The simple reason is that the two pans are different. One has polymer in it, and one doesn't. The polymer sample means there is extra material in the sample pan. Having extra material means that it will take more heat to keep the temperature of the sample pan increasing at the same rate as the reference pan.

So the heater underneath the sample pan has to work harder than the heater underneath the reference pan. It has to put out more heat. By measuring just *how much* more heat it has to put out is what we measure in a DSC experiment.

Specifically what we do is this: We make a plot as the temperature increases. On the *x*-axis we plot the temperature. On the *y*-axis we plot difference in heat output of the two heaters at a given temperature.

Heat Capacity

We can learn a lot from this plot. Let's imagine we're heating a polymer. When we start heating our two pans, the computer will plot the difference in heat output of the two heaters against temperature. That is to say, we're plotting the heat absorbed by the polymer against temperature. The plot will look something like this at first.

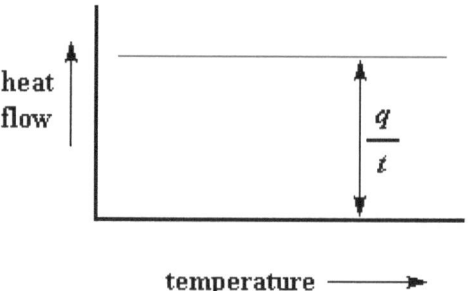

The heat flow at a given temperature can tell us something. The heat flow is going to be shown in units of heat, q supplied per unit time, t. The heating rate is temperature increase T per unit time, t. Got it?

$$\frac{\text{heat}}{\text{time}} = \frac{q}{t} = \text{heat flow}$$

$$\frac{\text{temperature increase}}{\text{time}} = \frac{\Delta T}{t} = \text{heating rate}$$

Let's say now that we divide the heat flow q/t by the heating rate T/t. We end up with heat supplied, divided by the temperature increase.

$$\frac{\dfrac{q}{t}}{\dfrac{\Delta T}{t}} = \frac{q}{\Delta T} = C_p = \text{heat capacity}$$

Remember from the glass transition page that when you put a certain amount of heat into something, its temperature will go up by a certain amount, and the amount of heat it takes to get a certain temperature increase is called the *heat capacity*, or C_p. We get the heat capacity by dividing the heat supplied by the resulting temperature increase. And that's just what we've done in that equation up there. We've figured up the heat capacity from the DSC plot.

The Glass Transition Temperature

Of course, we can learn a lot more than just a polymer's heat capacity with DSC. Let's see what happens when we heat the polymer a little more. After a certain temperature, our plot will shift upward suddenly, like this:

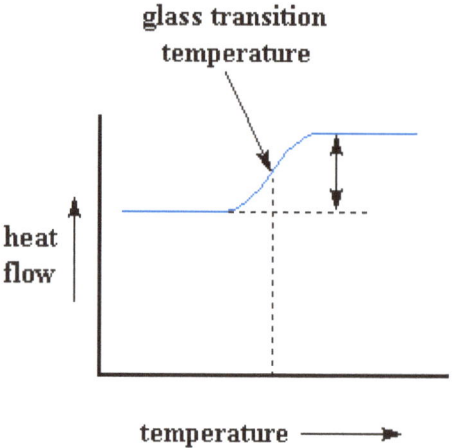

This means we're now getting more heat flow. This means we've also got an increase in the heat capacity of our polymer. This happens because the polymer has just gone through the glass transition. And as you learned on the glass transition page, polymers have a higher heat capacity above the glass transition temperature than they do below it. Because of this change in heat capacity that occurs at the glass transition, we can use DSC to measure a polymer's glass transition temperature. You may notice that the change doesn't occur suddenly, but takes place over a temperature range. This makes picking one discreet T_g kind of tricky, but we usually just take the middle of the incline to be the T_g.

Crystallization

But wait there is more, so much more. Above the glass transition, the polymers have a lot of mobility. They wiggle and squirm, and never stay in one position for very long. They're kind of like passengers trying to get comfortable in airline seats, and never quite succeeding, because they can move around more. When they reach the right temperature, they will have gained enough energy to move into very ordered arrangements, which we call crystals, of course.

When polymers fall into these crystalline arrangements, they give off heat. When this heat is dumped out, it makes the little computer-controlled heater under the sample pan really happy. It's happy because it doesn't have to put out much heat to keep the temperature of the sample pan rising. You can see this drop in the heat flow as a big dip in the plot of heat flow versus temperature:

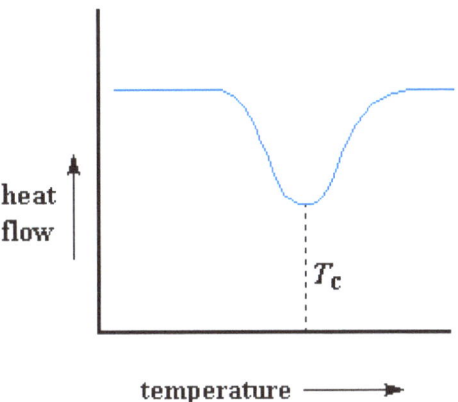

This dip tells us a lot of things. The temperature at the lowest point of the dip is usually considered to be the polymer's crystallization temperature, or T_c. Also, we can measure the area of the dip, and that will tell us the latent energy of crystallization for the polymer. But most importantly, this dip tells us that the polymer can in fact crystallize. If you analyzed a 100% amorphous polymer, like atactic polystyrene, you wouldn't get one of these dips, because such materials don't crystallize.

Also, because the polymer gives off heat when it crystallizes, we call crystallization an *exothermic* transition.

Melting

Heat may allow crystals to form in a polymer, but too much of it can be their undoing. If we keep heating our polymer past its T_c, eventually we'll reach another thermal transition, one called melting. When we reach the polymer's melting temperature, or T_m, those polymer crystals begin to fall apart, that is they melt. The chains come out of their ordered arrangements, and begin to move around freely. And in case you were wondering, we can spot this happening on a DSC plot.

Remember that heat that the polymer gave off when it crystallized? Well when we reach the T_m, it's payback time. There is a latent heat of melting as well as a latent heat of crystallization. When the polymer crystals melt, they must absorb heat in order to do so. Remember melting is a first order transition. This means that when you reach the melting temperature, the polymer's temperature won't rise until all the crystals have melted. This means that the little heater under the sample pan is going to have to put a lot of heat into the polymer in order to both melt the crystals *and* keep the temperature rising at the same rate as that of the reference pan. This extra heat flow during melting shows up as a big peak on our DSC plot, like this:

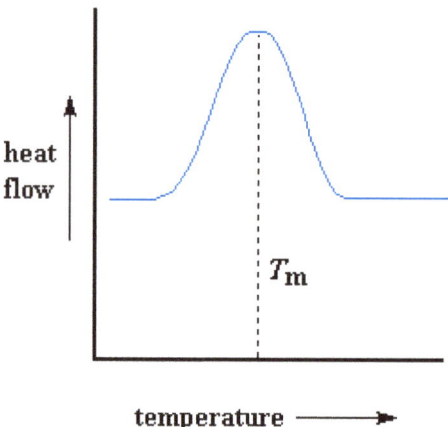

We can measure the latent heat of melting by measuring the area of this peak. And of course, we usually take the temperature at the top of the peak to be the polymer's melting temperature, T_m. Because we have to add energy to the polymer to make it melt, we call melting an *endothermic* transition.

Putting It All Together

So let's review now: we saw a step in the plot when the polymer was heated past its glass transition temperature. Then we saw a big dip when the polymer reached its crystallization temperature. Then finally we saw a big peak when the polymer reached its melting temperature. To put them all together, a whole plot will often look something like this:

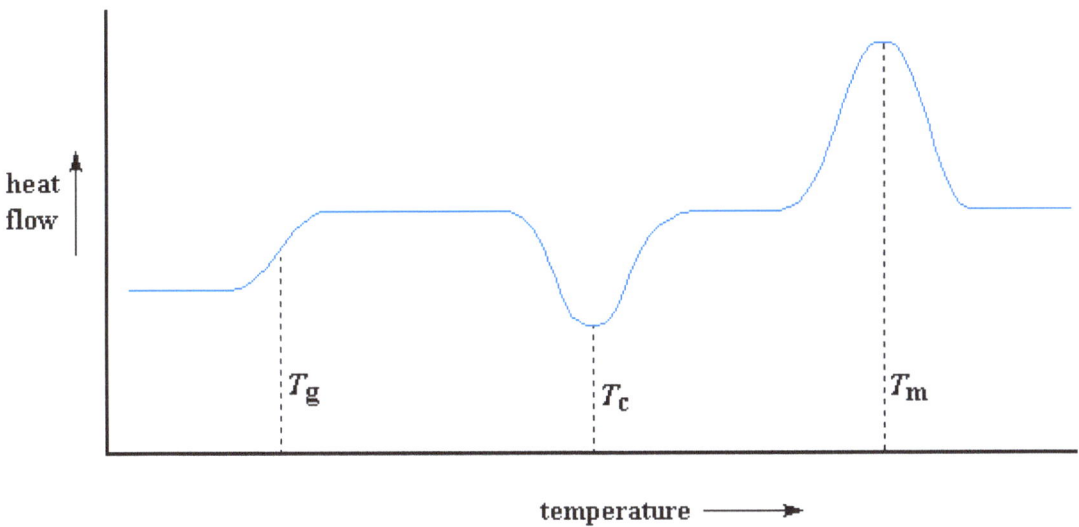

The entire DSC plot, right before your very eyes!

Of course, not everything you see here will be on every DSC plot. The crystallization dip and the melting peak will only show up for polymers that can form <u>crystals</u>. Completely amorphous polymers won't show any crystallization, or any melting either. But polymers with both crystalline and amorphous domains, will show all the features you see above.

If you look at the DSC plot you can see a big difference between the glass transition and the other two thermal transitions, crystallization and melting. For the glass transition, there is no dip, and there's no peak, either. This is because there is no latent heat given off, or absorbed, by the polymer during the glass transition. Both melting and crystallization involve giving off or absorbing heat. The only thing we do see at the glass transition temperature is a change in the heat capacity of the polymer.

Because there is a change in heat capacity, but there is no latent heat involved with the glass transition, we call the glass transition a *second order transition*. Transitions like melting and crystallization, which do have latent heats, are called *first order transitions*.

How much crystallinity?

DSC can also tell us how much of a polymer is crystalline and how much is amorphous. If you read the page dealing with polymer crystallinity, you know that many polymers contain both amorphous and crystalline material. But how much of each? DSC can tell us. If we know the latent heat of melting, ΔH_m, we can figure out the answer.

The first thing we have to do is measure the area of that big peak we have for the melting of the polymer. Now our plot is a plot of heat flow per gram of material, versus temperature. Heat flow is heat given off per second, so the area of the peak is given is units of heat x temperature x time^{-1} x mass^{-1}. We usually would put this in units such as joules x kelvins x (seconds)$^{-1}$ x (grams)$^{-1}$:

$$\text{area} = \frac{\text{heat} \times \text{temperature}}{\text{time} \times \text{mass}} = \frac{J\,K}{s\,g}$$

Got that? Don't worry. It gets simpler. We usually divide the area by the heating rate of our dsc experiment. The heating rate is in units of K/s. So the expression becomes simpler:

$$\frac{\text{area}}{\text{heating rate}} = \frac{\frac{J\,K}{s\,g}}{\frac{K}{s}} = \frac{J}{g}$$

Now we have a number of joules per gram. But because we know the mass of the sample, we can make it simpler. We just multiply this by the mass of the sample:

$$\left(\frac{J}{g}\right) \times g = J$$

Now we just calculated the total heat given off when the polymer melted. Neat, huh? Now if we do the same calculation for our dip that we got on the DSC plot for the crystallization of the polymer, we can get the total heat absorbed during the crystallization. We'll call the heat *total* heat given off during melting $H_{m, total}$, and we'll call the heat of the crystallization $H_{c, total}$.

Now we're going to subtract the two:

$$H_{m, total} - H_{c, total} = H'$$

Why did we just do that? And what does that number H' mean? H' is the heat given off by that part of the polymer sample which was already in the crystalline state *before* we heated the polymer above the T_c. We want to know how much of the polymer was crystalline before we induced more of it to become crystalline. That's why we subtract the heat given off at crystallization. Is everyone following me?

Now with our magic number H' we can figure up the percent crystallinity. We're going to divide it by the specific heat of melting, H_c^*. The specific heat of melting? That's the amount of heat given off by a certain amount, usually one gram, of a polymer. H' is in joules, and the specific heat of melting is usually given in joules per gram, so we're going to get an answer in grams, which we'll call m_c.

$$\frac{H'}{H^*_m} = m_c \qquad \frac{\dfrac{J}{J}}{g} = g$$

This is the total amount of grams of polymer that were crystalline below the T_c. Now if we divide this number by the weight of our sample, m_{total}, we get the fraction of the sample that was crystalline, and then of course, the percent crystallinity:

$$\frac{m_c}{m_{total}} = \text{crystalline fraction}$$

$$\text{crystalline fraction} \times 100 = \% \text{ crystallinity}$$

Applications

Differential scanning calorimetry can be used to measure a number of characteristic properties of a sample. Using this technique it is possible to observe fusion and crystallization events as well as glass transition temperatures T_g. DSC can also be used to study oxidation, as well as other chemical reactions.

Glass transitions may occur as the temperature of an amorphous solid is increased. These transitions appear as a step in the baseline of the recorded DSC signal. This is due to the sample undergoing a change in heat capacity; no formal phase change occurs.

As the temperature increases, an amorphous solid will become less viscous. At some point the molecules may obtain enough freedom of motion to spontaneously arrange themselves into a crystalline form. This is known as the crystallization temperature (T_c). This transition from amorphous solid to crystalline solid is an exothermic process, and results in a peak in the DSC signal. As the temperature increases the sample eventually reaches its melting temperature (T_m). The melting process results in an endothermic peak in the

DSC curve. The ability to determine transition temperatures and enthalpies makes DSC a valuable tool in producing phase diagrams for various chemical systems.

DSC is used widely for examining polymeric materials to determine their thermal transitions. The observed thermal transitions can be utilized to compare materials, however, the transitions do not uniquely identify composition. Composition of unknown materials may be completed using a technique such as IR. Melting points and glass transition temperatures for most polymers are available from standard compilations, and the method can show polymer degradation by the lowering of the expected melting point, T_m, for example. T_m depends on the molecular weight of the polymer and thermal history, so lower grades may have lower melting points than expected. The percent crystalline content of a polymer can be estimated from the crystallization/melting peaks of the DSC graph as reference heats of fusion can be found in the literature. DSC can also be used to study thermal degradation of polymers using an approach such as Oxidative Onset Temperature/Time (OOT), however, the user risks contamination of the DSC cell, which can be problematic. Thermogravimetric Analysis (TGA) may be more useful for decomposition behavior determination. Impurities in polymers can be determined by examining thermograms for anomalous peaks, and plasticisers can be detected at their characteristic boiling points. In addition, examination of minor events in first heat thermal analysis data can be useful as these apparently "anomalous peaks" can in fact also be representative of process or storage thermal history of the material or polymer physical aging. Comparison of first and second heat data collected at consistent heating rates can allow the analyst to learn about both polymer processing history and material properties.

DSC is used in the study of liquid crystals. As some forms of matter go from solid to liquid they go through a third state, which displays properties of both phases. This anisotropic liquid is known as a liquid crystalline or mesomorphous state. Using DSC, it is possible to observe the small energy changes that occur as matter transitions from a solid to a liquid crystal and from a liquid crystal to an isotropic liquid.

DSC is widely used in the pharmaceutical and polymer industries. For the polymer chemist, DSC is a handy tool for studying curing processes, which allows the fine tuning of polymer properties. The cross-linking of polymer molecules that occurs in the curing process is exothermic, resulting in a positive peak in the DSC curve that usually appears soon after the glass transition. In the pharmaceutical industry it is necessary to have well-characterized drug compounds in order to define processing parameters. For instance, if it is necessary to deliver a drug in the amorphous form, it is desirable to process the drug at temperatures below those at which crystallization can occur

Chapter 4 ELISA

Introduction

The enzyme-linked immunosorbent assay (ELISA) is a test that uses antibodies and color change to identify a substance.

ELISA is a popular format of a "wet-lab" type analytic biochemistry assay that uses a solid-phase enzyme immunoassay (EIA) to detect the presence of a substance, usually an antigen, in a liquid sample or wet sample.

The ELISA has been used as a diagnostic tool in medicine and plant pathology, as well as a quality-control check in various industries.

Antigens from the sample are attached to a surface. Then, a further specific antibody is applied over the surface so it can bind to the antigen. This antibody is linked to an enzyme, and, in the final step, a substance containing the enzyme's substrate is added. The subsequent reaction produces a detectable signal, most commonly a color change in the substrate.

Performing an ELISA involves at least one antibody with specificity for a particular antigen. The sample with an unknown amount of antigen is immobilized on a solid support (usually a polystyrene microtiter plate) either non-specifically (via adsorption to the surface) or specifically (via capture by another antibody specific to the same antigen, in a "sandwich" ELISA). After the antigen is immobilized, the detection antibody is added, forming a complex with the antigen. The detection antibody can be covalently linked to an enzyme, or can itself be detected by a secondary antibody that is linked to an enzyme through bioconjugation. Between each step, the plate is typically washed with a mild detergent solution to remove any proteins or antibodies that are not specifically bound. After the final wash step, the plate is developed by adding an enzymatic substrate to produce a visiblesignal, which indicates the quantity of antigen in the sample.

Of note, ELISA can perform other forms of ligand binding assays instead of strictly "immuno" assays, though the name carried the original "immuno" because of the common use and history of development of this method. The technique essentially requires any ligating reagent that can be immobilized on the solid phase along with a detection reagent that will bind specifically and use an enzyme to generate a signal that can be properly quantified. In between the washes, only the ligand and its specific binding counterparts remain specifically bound or "immunosorbed" by antigen-antibody interactions to the solid phase, while the nonspecific or unbound components are washed away. Unlike other spectrophotometric wet lab assay formats where the same reaction well (e.g. a cuvette) can be reused after washing, the ELISA plates have the reaction products immunosorbed on the solid phase which is part of the plate, so are not easily reusable.

Principle

As a "WET LAB" analytic biochemistry assay, ELISA involves detection of an "analyte" (i.e. the specific substance whose presence is being quantitatively or qualitatively analyzed) in a liquid sample by a method that continues to use liquid reagents during the "analysis" (i.e. controlled sequence of biochemical reactions that will generate a signal which can be easily quantified and interpreted as a measure of the amount of analyte in the sample) that stays liquid and remains inside a reaction chamber or well needed to keep the reactants contained; It is opposed to "dry lab" that can use dry strips - and even if the sample is liquid (e.g. a measured small drop), the final detection step in "dry" analysis involves reading of a dried strip by methods such as reflectometryand does not need a reaction containment chamber to prevent spillover or mixing between samples.

As a heterogenous assay, ELISA separates some component of the analytical reaction mixture by adsorbing certain components onto a solid phase which is physically immobilized. In ELISA, a liquid sample is added onto a stationary solid phase with special binding properties and is followed by multiple liquid reagents that are sequentially added, incubated and washed followed by some optical change (e.g. color development by the product of an enzymatic reaction) in the final liquid in the well from which the quantity of the analyte is measured. The qualitative "reading" usually based on detection of intensity of transmitted light by spectrophotometry, which involves quantitation of transmission of some specific wavelength of light through the liquid (as well as the transparent bottom of the well in the multiple-well plate format). The sensitivity of detection depends on amplification of the signal during the analytic reactions. Since enzyme reactions are very well known amplification processes, the signal is generated by enzymes which are linked to the detection reagents in fixed proportions to allow accurate quantification - thus the name "enzyme linked".

The analyte is also called the ligand because it will specifically bind or ligate to a detection reagent, thus ELISA falls under the bigger category of ligand binding assays. The ligand-specific binding reagent is "immobilized", i.e., usually coated and dried onto the transparent bottom and sometimes also side wall of a well (the stationary "solid phase'/"solid substrate" here as opposed to solid microparticle/beads that can be washed away), which is usually constructed as a multiple-well plate known as the "ELISA plate". Conventionally, like other forms of immunoassays, the specificity of antigen-antibody type reaction is used because it is easy to raise an antibody specifically against an antigen in bulk as a reagent. Alternatively, if the analyte itself is an antibody, its target antigen can be used as the binding reagent

Types of ELISA

Indirect ELISA

The steps of "indirect" ELISA follow the mechanism below:-

- A buffered solution of the antigen to be tested for is added to each well of a microtiter plate, where it is given time to adhere to the plastic through charge interactions.

- A solution of nonreacting protein, such as bovine serum albumin or casein, is added to well (usually 96-well plates) any plastic surface in the well that remains uncoated by the antigen.

- The primary antibody is added, which binds specifically to the test antigen coating the well. This primary antibody could also be in the serum of a donor to be tested for reactivity towards the antigen.

- A secondary antibody is added, which will bind the primary antibody. This secondary antibody often has an enzyme attached to it, which has a negligible effect on the binding properties of the antibody. In other cases, as in the diagram to the left, the primary antibody itself is conjugated to the enzyme.

- A substrate for this enzyme is then added. Often, this substrate changes color upon reaction with the enzyme. The color change shows the secondary antibody has bound to primary antibody, which strongly implies the donor has had an immune reaction to the test antigen. This can be helpful in a clinical setting, and in research.

- The higher the concentration of the primary antibody present in the serum, the stronger the color change. Often, a spectrometer is used to give quantitative values for color strength.

The enzyme acts as an amplifier; even if only few enzyme-linked antibodies remain bound, the enzyme molecules will produce many signal molecules.

Within common-sense limitations, the enzyme can go on producing color indefinitely, but the more primary antibody is present in the donor serum, the more secondary antibody + enzyme will bind, and the faster the color will develop. A major disadvantage of the indirect ELISA is the method of antigen immobilization is not specific; when serum is used as the source of test antigen, all proteins in the sample may stick to the microtiter plate well, so small concentrations of analyte in serum must compete with other serum proteins when binding to the well surface. The sandwich or direct ELISA provides a solution to this problem, by using a "capture" antibody specific for the test antigen to pull it out of the serum's molecular mixture.

ELISA may be run in a qualitative or quantitative format. Qualitative results provide a simple positive or negative result (yes or no) for a sample. The cutoff between positive and negative is determined by the analyst and may be statistical. Two or three times the standard deviation (error inherent in a test) is often used to distinguish positive from negative samples. In quantitative ELISA, the optical density (OD) of the sample is compared to a standard curve, which is typically a serial dilution of a known-concentration solution of the target molecule. For example, if a test sample returns an OD of 1.0, the point on the standard curve that gave OD = 1.0 must be of the same analyte concentration as the sample.

Figure 14 Indirect ELISA

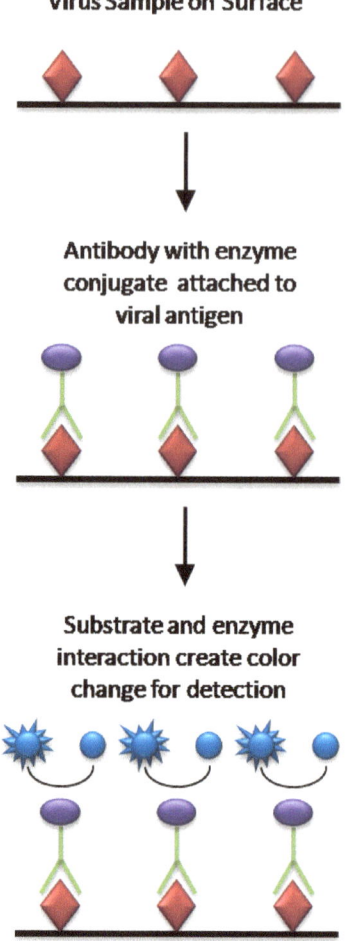

Virus Sample on Surface

Antibody with enzyme conjugate attached to viral antigen

Substrate and enzyme interaction create color change for detection

Sandwich ELISA

A "sandwich" ELISA, is used to detect sample antigen. The steps are:

1. A surface is prepared to which a known quantity of capture antibody is bound.

2. Any nonspecific binding sites on the surface are blocked.

3. The antigen-containing sample is applied to the plate.

4. The plate is washed to remove unbound antigen.

5. A specific antibody is added, and binds to antigen (hence the 'sandwich': the Ag is stuck between two antibodies)

6. Enzyme-linked secondary antibodies are applied as detection antibodies that also bind specifically to the antibody's Fc region (nonspecific).

7. The plate is washed to remove the unbound antibody-enzyme conjugates.

8. A chemical is added to be converted by the enzyme into a color or fluorescent or electrochemical signal.

9. The absorbency or fluorescence or electrochemical signal (e.g., current) of the plate wells is measured to determine the presence and quantity of antigen.

The image to the right includes the use of a secondary antibody conjugated to an enzyme, though, in the technical sense, this is not necessary if the primary antibody is conjugated to an enzyme. However, use of a secondary-antibody conjugate avoids the expensive process of creating enzyme-linked antibodies for every antigen one might want to detect. By using an enzyme-linked antibody that binds the Fc region of other antibodies, this same enzyme-linked antibody can be used in a variety of situations. Without the first layer of "capture" antibody, any proteins in the sample (including serum proteins) may competitively adsorb to the plate surface, lowering the quantity of antigen immobilized. Use of the purified specific antibody to attach the antigen to the plastic eliminates a need to purify the antigen from complicated mixtures before the measurement, simplifying the assay, and increasing the specificity and the sensitivity of the assay.

Figure 15   Sandwich ELISA

(1)   (2)   (3)   (4)   (5)

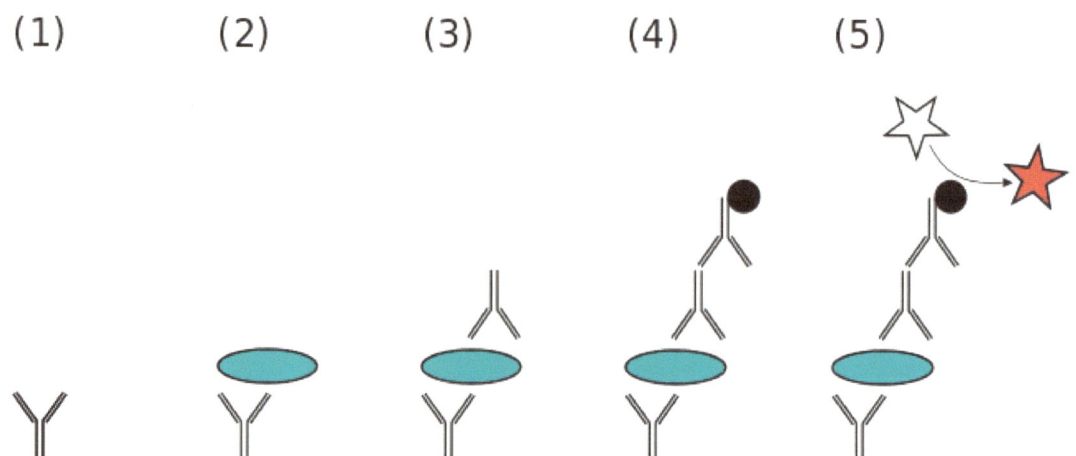

Competitive ELISA

A third use of ELISA is through competitive binding. The steps for this ELISA are somewhat different from the first two examples:

1. Unlabeled antibody is incubated in the presence of its antigen (sample).

2. These bound antibody/antigen complexes are then added to an antigen-coated well.

3. The plate is washed, so unbound antibody is removed. (The more antigen in the sample, the less antibody will be able to bind to the antigen in the well, hence "competition".)

4. The secondary antibody, specific to the primary antibody, is added. This second antibody is coupled to the enzyme.

5. A substrate is added, and remaining enzymes elicit a chromogenic or fluorescent signal.

6. The reaction is stopped to prevent eventual saturation of the signal.

Some competitive ELISA kits include enzyme-linked antigen rather than enzyme-linked antibody. The labeled antigen competes for primary antibody binding sites with the sample antigen (unlabeled). The less antigen in the sample, the more labeled antigen is retained in the well and the stronger the signal.

Figure 16 Competitive ELISA

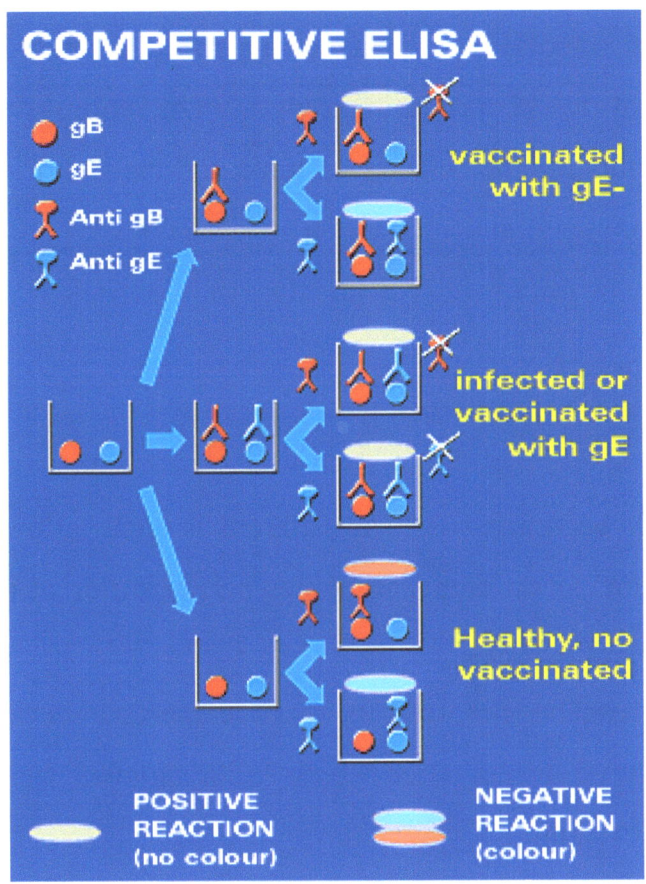

Commonly, the antigen is not first positioned in the well.

For the detection of HIV antibodies, the wells of microtiter plate are coated with the HIV antigen. Two specific antibodies are used, one conjugated with enzyme and the other present in serum (if serum is positive for the antibody). Cumulative competition occurs between the two antibodies for the same antigen, causing a stronger signal to be seen. Sera to be tested are added to these wells and incubated at 37°C, and then washed. If antibodies are present, the antigen-antibody reaction occurs. No antigen is left for the enzyme-labelled specific HIV antibodies. These antibodies remain free upon addition and are washed off during washing. Substrate is added, but there is no enzyme to act on it, so positive result shows no color change.

Multiple and portable ELISA

A new technique (EP 1 499 894 B1 in EPO Bulletin 25.02.209 N. 2009/09; USPTO 7510687 in USPTO Bulletin 31.03.2009; ZL 03810029.0 in SIPO PRC Bulletin 08.04.2009) uses a solid phase made up of an immunosorbent polystyrene rod with eight to 12 protruding ogives. The entire device is immersed in a test tube containing the collected sample and the following steps (washing, incubation in conjugate and incubation in chromogens) are carried out by dipping the ogives in microwells of standard microplates filled with reagents.

The advantages of this technique are:

1. The ogives can each be sensitized to a different reagent, allowing the simultaneous detection of different antibodies and/or different antigens for multiple-target assays.
2. The sample volume can be increased to improve the test sensitivity in clinical (blood, saliva, urine), food (bulk milk, pooled eggs) and environmental (water) samples.
3. One ogive is left unsensitized to measure the nonspecific reactions of the sample.
4. The use of laboratory supplies for dispensing sample aliquots, washing solution and reagents in microwells is not required, facilitating the development of ready-to-use lab kits and on-site testing.

ELISA can be performed to evaluate either the presence of antigen or the presence of antibody in a sample, it is a useful tool for determining serum antibody concentrations (such as with the HIV test] or West

Nile virus). It has also found applications in the food industry in detecting potential food allergens, such as milk, peanuts, walnuts, almonds, and eggs. ELISA can also be used in toxicology as a rapid presumptive screen for certain classes of drugs.

Application

The ELISA was the first screening test widely used for HIV because of its high sensitivity. In an ELISA, a person's serum is diluted 400 times and applied to a plate to which HIV antigens are attached. If antibodies to HIV are present in the serum, they may bind to these HIV antigens. The plate is then washed to remove all other components of the serum. A specially prepared "secondary antibody" — an antibody that binds to other antibodies — is then applied to the plate, followed by another wash. This secondary antibody is chemically linked in advance to an enzyme.

Thus, the plate will contain enzyme in proportion to the amount of secondary antibody bound to the plate. A substrate for the enzyme is applied, and catalysis by the enzyme leads to a change in color or fluorescence. ELISA results are reported as a number; the most controversial aspect of this test is determining the "cut-off" point between a positive and a negative result.

A cut-off point may be determined by comparing it with a known standard. If an ELISA test is used for drug screening at workplace, a cut-off concentration, 50 ng/ml, for example, is established, and a sample containing the standard concentration of analyte will be prepared. Unknowns that generate a stronger signal than the known sample are "positive." Those that generate weaker signal are "negative".

Dr Dennis E Bidwell and Alister Voller created the ELISA test to detect various kind of diseases, such as malaria, Chagas disease, and Johne's disease ELISA tests also are used as in *in vitro* diagnostics in medical laboratories. The other uses of ELISA include:

- detection of *Mycobacterium* antibodies in tuberculosis
- detection of rotavirus in feces
- detection of hepatitis B markers in serum
- detection of enterotoxin of *E. coli* in feces
- detection of HIV antibodies in blood samples

Chapter 5 Radioimmunoassay (RIA)

INTRODUCTION

Radioimmunoassay (RIA) is a nuclear technique widely used for measuring minute substances with in vitro procedures. It combined the technology of nuclear medicine (tracer technique) and immunology (antigen-antibody binding) so that the name RIA is designated.

This technique has been developed in 1959 by S. Berson and R. Yalow. They measured insulin in the serum. For her contribution of this important analytical technique to medical science, R. Yalow shared the 1977 Nobel price in medicine and physiology.

In earlier days RIA only limited to certain hormone, but now the scope greatly expanded to the field of reproductive physiology, oncology, immunology, hematology, pharmacology and parasitology etc. It makes a great contribution to the diagnostic laboratory and scientific research works.

RIA is a sensitive, specific, precise and accurate technique, and a in vitro procedure without pain. But it maybe cause radioactive contamination. The equipment is expensive.

PRINCIPLE OF RIA

There are three major substances in the reactive system: antigen (Ag), labeled antigen (*Ag) and antibody (Ab). Ag is unknown, to be assayed in serum, or the standard Ag, which is variant. *Ag is the same antigen added in minute amount of radioactivity. It is constant. Ab can specially bind to the Ag and *Ag. It is constant and limited.

Labeled antigen (*Ag) possesses the same properties of unlabeled antigen (Ag). It can also bind to the correlated specific antibody (Ab) with the formation of labeled antigen-antibody complex or called bound antigen (B), leave the unbound one as free labeled antigen (F). The more Ag is present, the less likely is the *Ag bound to the Ab, thus the amount of B formed is inversely proportional to the Ag originally present in serum, this is so called competing depression reaction.

BASIC TECHNIQUES

RIA must have four basic techniques, they are standard substance, labeled antigen, specific binding substance, separating reagent. (1) Standard substance is the fundamental of accurate quantitation, is regarded as the poise. It must be as same as the antigen which will be assayed, and with high purity. The quantity of the standard substance has to be precision. (2) There are only minute amount of labeled antigen that is less than the minimum of the antigen to be assayed. Labeled antigen competes with assayed antigen (or standard substance) to bind to the antibody. It must have optimal specific activity, high radiochemical purity, good immunoactivity and stability. (3) Antibody become the most common specific binding substance for its high binding affinity. The higher affinity antibody has, the more sensitive we get. Besides antibody, there are some specific binding globulin, such as cortical binding globulin for cortisol, thyroid binding globulin for thyroid hormone, sex binding globulin for progestin and estrogen, etc. (4) Separation of B and F must be complete, so the separating reagent is very important. We usually use the following three kinds of separating reagent. First is nonspecific separating reagent which can absorb F. The next is specific separating reagent, such as second antibody that can bind to antibody. The last is solid separating reagent which absorb antibody on solid material

Figure 17 Radioimmunoassay (RIA)

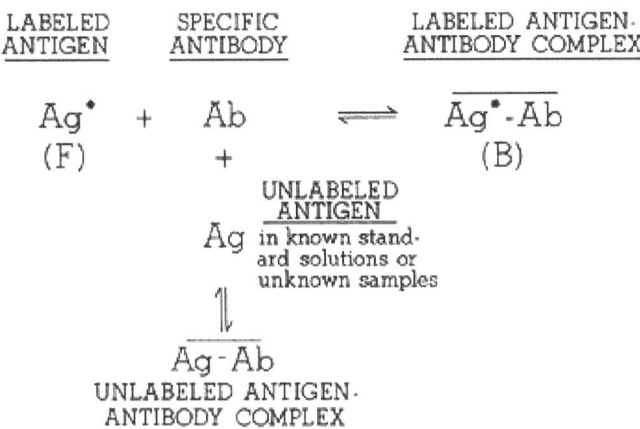

RIA Tracers

1. *Internally* labeled molecules

- Typically, **tritium (3H)** is the label, sometimes 14C. Usually purchased commercially. Used only for small molecules like steroids or drugs.

- Pro : the tracer immunologically behaves exactly as the cold hormone, thus theoretically perfect.

- Con : beta-radiation is weak and therefore more difficult to measure, thus practically cumbersome.

- Beta rays have low penetrating power: the radioactive sample needs to be mixed with a scintillator fluid; the produced light is measured by use of a photomultiplier ("beta-counter").

2. *Externally* labeled molecules

- Typically, **125I** is the label.

- Pro : often produced in the research lab itself.

- Pro : gamma-radiation has high penetrating power, is therefore easy to measure, thus practical to use.
- Con : the tracer immunologically does *not* always behave exactly as the cold hormone, due to iodination damage.
- Con : shelf-life of iodinated protein is < 4 weeks.
- Usually, 125I will take the place of a hydrogen atom on on the ring of tyrosine.
- Sometimes, a radioactively labeled molecule needs to be conjugated to the protein.

Advantages

- versatility : using the same principle, almost any biomolecule can be assayed
- fast (usually 2 days or less)
- sensitive (comparable to the most sensitive bioassays, that is < ng/ml)
- large capacity : thousands of samples/day
- specific (antibody-dependent)

Disadvantages

- Use of radioactivity: hazardous
- Expensive equipment (gamma or beta counter)